CLIMATE

THE EFFECTS OF CLIMATE CHANGE

BY MARTHA LONDON

CONTENT CONSULTANT
Atreyee Bhattacharya, PhD
Research Affiliate, Institute of Arctic and Alpine Research
University of Colorado Boulder;
Visiting Researcher, Scripps Institution of Oceanography
University of California-San Diego

Cover image: Climate change is causing habitat loss for polar bears.

Core Library
An Imprint of Abdo Publishing
abdobooks.com

abdobooks.com

Published by Abdo Publishing, a division of ABDO, PO Box 398166, Minneapolis, Minnesota 55439. Copyright © 2021 by Abdo Consulting Group, Inc. International copyrights reserved in all countries. No part of this book may be reproduced in any form without written permission from the publisher. Core Library™ is a trademark and logo of Abdo Publishing.

Printed in the United States of America, North Mankato, Minnesota
082020
012021

THIS BOOK CONTAINS RECYCLED MATERIALS

Cover Photo: Kotomiti Okuma/Shutterstock Images
Interior Photos: Samuel Boivin/NurPhoto/Getty Images, 4–5; Sylvain Thomas/AFP/Getty Images, 6; Red Line Editorial, 9; iStockphoto, 12–13, 29, 35 (hurricane symbol); USGS/Science Source, 16 (top); USGS/Karen Holzer/Science Source, 16 (bottom); Fortgens Photography/iStockphoto, 18–19; Timothy Epp/Shutterstock Images, 20, 45; Shutterstock Images, 22–23, 32–33, 43; Bryan and Cherry Alexander/Science Source, 24; Geoffrey Reynaud/iStockphoto, 25; Elliotte Rusty Harold/Shutterstock Images, 27; Arthur G. Photography/Shutterstock Images, 37; Pedro Sala/iStockphoto, 39

Editor: Marie Pearson
Series Designer: Katharine Hale

Library of Congress Control Number: 2019954197

Publisher's Cataloging-in-Publication Data

Names: London, Martha, author
Title: The effects of climate change / by Martha London
Description: Minneapolis, Minnesota : Abdo Publishing, 2021 | Series: Climate change | Includes online resources and index.
Identifiers: ISBN 9781532192739 (lib. bdg.) | ISBN 9781644944264 (pbk.) | ISBN 9781098210632 (ebook)
Subjects: LCSH: Climatic changes--Social aspects--Juvenile literature. | Global temperature changes--Juvenile literature. | Climatic changes--Economic aspects--Juvenile literature. | Environmental effects on human beings--Juvenile literature. | Atmospheric greenhouse effect--Juvenile literature.
Classification: DDC 363.738--dc23

CONTENTS

CHAPTER ONE
Rising Temperatures 4

CHAPTER TWO
Melting Ice 12

CHAPTER THREE
Effects on Plants and Animals 22

CHAPTER FOUR
Stronger Storms 32

Fast Facts 42

Stop and Think 44

Glossary 46

Online Resources 47

Learn More 47

Index 48

About the Author 48

CHAPTER ONE

RISING TEMPERATURES

A woman in France lay next to a fountain. The air temperature was more than 110 degrees Fahrenheit (43°C). Sweat dripped down her cheeks. Spray from the fountain cooled her briefly.

Children at a school fanned themselves. They were glad they did not have to take their tests. It was too hot. On any other day, the kids might have wanted to play outside. But it was too hot to do anything. The heat was making people sick. Some people had even died.

People in France tried to cool off in pools and fountains during the 2019 heat wave.

During the 2019 heat wave in Europe, wildlife centers received animals that were struggling due to a lack of food and water.

The woman at the fountain worried about her parents. Elderly people have the most trouble in the heat. The heat makes it more difficult to breathe. Most European homes do not have air conditioning. Europe is not usually this hot. That night a news station reported on a wildfire in Spain. High temperatures can make land drier, making fires more likely. Heat warnings were in place throughout Europe.

The woman, the school kids, and many other Europeans waited for the weather to cool. It eventually did. But scientists say that heat waves like this one, which happened in Europe in July 2019, will become more common. Warmer temperatures are a direct result of climate change.

WHAT'S CAUSING THE WARMING?

Much of the world uses fossil fuels. People burn fossil fuels to heat homes and power transportation. People burn these fuels at power

PERSPECTIVES

AIR QUALITY IMPACT

Climate change can make air quality worse than usual. In 2016 the US government released a report on the effects of air quality on human health. Officials found that air quality will get worse over time. For example, drier conditions cause more forest fires. Forest fires release particles into the air. These particles can cause breathing trouble. Poor air quality is bad for people who have allergies or asthma. In the 2016 report, the Environmental Protection Agency explained that climate change can cause poor air quality in both outdoor and indoor air.

plants to generate electricity. Fossil fuels are resources such as coal, oil, and natural gas. They take millions of years to form. Fossil fuels are not renewable. This means they cannot be used over and over without being used up. People burn oil faster than it is created naturally.

Burning oil and coal creates carbon dioxide. Carbon dioxide is a greenhouse gas. Greenhouse gases trap heat in the atmosphere. They keep heat from escaping into space. This is how Earth stays warm. Having some greenhouse gases keeps the planet warm enough for life to survive. But adding too much of these gases traps even more heat. Earth's temperature rises quickly. This is called global warming.

EFFECTS OF RISING TEMPERATURES

A warming Earth causes glaciers and sea ice to melt. This includes the Arctic and Antarctic ice caps. Glaciers are disappearing even faster than scientists predicted. Researchers are worried this means climate change is also speeding up.

GLOBAL TEMPERATURE CHANGE

Between 1951 and 1980, scientists calculated the average global temperature for all 30 years. This temperature serves as the 0 on the above graph. The graph shows how the average yearly temperature changed from 1880 to 2018, compared to that average. What are two things you notice about the graph? How does the graph help you better understand the information in Chapter One?

While glaciers melt, high temperatures cause other areas to dry out. This creates drought conditions around the world. Areas in Asia, Latin America, and Africa are especially affected. Many of these areas do not have enough water to drink or grow crops. Wildfires become more common where there is drought. In the United States, California has seen more wildfires

> **DROUGHT IN AFRICA**
>
> Scientists estimate that 300 million people in Africa use corn for food. However, climate change is making droughts worse in Africa. People struggle to grow corn. Farmers rely on rain. There are few sources of water for their crops. Without rain, people are struggling to grow enough food.

each year. The ground is dry. Fires spread easily. From 2019 to 2020, wildfires raged through Australia. Hot temperatures and dry conditions made the fires especially intense.

Other areas are seeing more rain than normal. Flooding is happening more often. For example, in 2019, several US states in the Midwest flooded. Rivers such as the Mississippi flowed over their banks. It took several weeks for the waters to go down. Farmers in Iowa and Missouri lost many of their crops.

Countries around the world are seeing the effects of climate change. Rising temperatures create conditions that set off a chain reaction. Climate change affects every part of the world.

STRAIGHT TO THE SOURCE

In 2019 climate activist Greta Thunberg was 16 years old. Greta asked politicians to change laws about fossil fuels. That year she gave a speech in front of the British government. She said:

> Around the year 2030, . . . we will be in a position where we set off an irreversible chain reaction beyond human control, that will most likely lead to the end of our [civilization] as we know it. That is unless in that time, permanent and unprecedented changes in all aspects of society have taken place, including a reduction of [carbon dioxide] emissions by at least 50%.
>
> And please note that these calculations are depending on inventions that have not yet been invented at scale, inventions that are supposed to clear the atmosphere of astronomical amounts of carbon dioxide.

Source: Greta Thunberg. "You Did Not Act in Time." *Guardian*, 23 Apr. 2019, theguardian.com. Accessed 31 Oct. 2019.

BACK IT UP

The speaker is using evidence to support a point. Write a paragraph describing the point she is making. Then write down two or three pieces of evidence she uses to make the point.

CHAPTER TWO

MELTING ICE

There are two kinds of ice. There is sea ice, and there is land ice. Sea ice is found in the oceans around the Arctic and Antarctic. Land ice includes glaciers on land. Sea ice is made from salt water, though the ice contains very little salt. Land ice forms from freshwater sources, such as snow. Areas with large amounts of land ice include Greenland, the Andes Mountains in South America, and Glacier National Park in the United States. Land ice stays year after year. Some melting

Ice can form in sheets on top of seawater.

occurs every year. But historically, new ice and melted ice balanced each other out. That is not the case today.

Ice around the globe is melting. Much of the land and sea ice that exists has been present for thousands or millions of years. As ice melts, there are several consequences. Melting ice affects nations around the world.

RISING OCEANS

As climate warming continues, ice melts faster than it forms. When sea ice melts, the

PERSPECTIVES

THE FIRST LOST GLACIER

In 2014 a glacier in Iceland nicknamed Ok melted so much that it could not be considered a glacier anymore. The melting was caused by climate change. In 2019 scientists came together. They created a plaque to remember the lost glacier. The text on the plaque reads: "Ok is the first Icelandic glacier to lose its status as a glacier. In the next 200 years, all our glaciers are expected to follow the same path. This monument is to acknowledge that we know what is happening and what needs to be done. Only you know if we did it."

water levels stay the same. The reason for this is that sea ice is made from frozen seawater. Sea ice is already in the ocean. It is similar to ice in a glass of water. Melted ice cubes do not cause the glass to overflow.

In contrast, melting land ice is what causes sea levels to rise. Land ice trapped in glaciers is fresh water. As it melts, it flows downhill. Some of the water is absorbed by the land. Some flows into rivers and lakes. Eventually, the melted water arrives at the ocean. This adds new water to the ocean. The sea level rises. Sea levels have risen as much as 8 inches (20 cm) in the last 100 years.

Such a dramatic change in sea level has a big effect on small islands. In 2017 scientists found that at least eight islands in the Pacific Ocean had disappeared. The islands were small and had few trees. Ocean waves washed sand away, and rising sea levels submerged the islands.

The lost islands were part of a region called Oceania. Oceania contains thousands of small island nations. Some nations are making evacuation plans. If sea levels continue to rise, people will not be able to continue living on some of the islands.

Rising sea levels aren't the only danger caused by melting land ice. Melted ice flows into rivers. The rivers swell over their banks. This swelling can happen quickly, causing a flash flood. Flash floods destroy trees, homes, and even whole communities.

ISLANDS IN THE ARCTIC

Arctic ice in Canada has been melting since the 1960s. That melting began to speed up in the 1990s. In 2017 scientists were startled by a satellite image. Three new islands in the Canadian Arctic sat in the ocean. The islands used to be covered by ice. But the ice melted and retreated, exposing the islands.

A photo from the 1940s, *top*, shows the Grinnell Glacier at Glacier National Park. A photo from 2006, *bottom*, shows how much it has melted. The glacier is still melting today.

A glacier once filled this mountain valley in Austria. It has been melting since 1850, exposing the dark ground beneath.

DARK ROCK, DARK OCEAN

Ice acts as a natural cooling system for Earth. When sunlight hits the ground, it produces warmth. White ice and snow reflect sunlight back into the atmosphere. Some of that energy escapes into space. As glaciers and ice sheets melt, there is less snow and ice to reflect sunlight.

Beneath the layers of sea ice are rock and ocean. Seawater and rock are dark in color. Dark colors

absorb sunlight. They heat up. Warm rock and water speed up melting. This is called a feedback loop. As more ice melts, more water and rock are exposed. They absorb more heat from sunlight, causing more ice to melt. Feedback loops are difficult to stop.

As water gets warmer, it expands. Rising sea levels are due in part to warming ocean temperatures in addition to melting land ice. Scientists say that warmer oceans are speeding up the process of rising sea levels.

Melting permafrost can cause buildings to shift and eventually collapse.

MELTING PERMAFROST

A lot of ice exists just beneath Earth's surface. In Arctic regions, a layer of dirt is permanently frozen. It is called permafrost. People build houses into the permafrost layer because it is stable. Its name suggests it will never melt. However, climate change is causing permafrost in some areas to melt.

Ice takes up more space than liquid water. When the ice in the soil melts, much of the land sinks. In Canada and Alaska, roads that used to be flat now have hills. Houses and buildings have become unsafe to live in. Without the hard layer of permafrost, buildings are in danger of collapsing.

Permafrost holds carbon and another greenhouse gas called methane. As permafrost melts, it releases methane and carbon. These gases build up in the atmosphere, trapping more heat. As with dark rock and seawater, permafrost melt causes a feedback loop.

EXPLORE ONLINE

Chapter Two focuses on the effects melting glaciers have on the planet. One of the biggest effects is rising sea levels. The website below offers more information about the effects of rising sea levels. How is the information on the website similar to the information in Chapter Two? What new information did you learn from the website?

SEA LEVEL RISE

abdocorelibrary.com/effects-of-climate-change

CHAPTER THREE

EFFECTS ON PLANTS AND ANIMALS

Climate change has a big effect on Earth's ecosystems. As habitats get warmer, wetter, or drier, plants and animals have to adapt. But changes are happening quickly. Some plants and animals are losing their habitats faster than they can adapt.

SHRINKING HABITATS

The Arctic is warming faster than other places on Earth. As a result, sea ice is disappearing.

Polar bears are one of the millions of species threatened by climate change.

Less sea ice puts seals at risk because they need sea ice to raise their pups.

Both warming temperatures and less sea ice affect Arctic animals such as polar bears and caribou.

Polar bears and other polar species are especially affected by the disappearing ice. Polar bears need sea ice to hunt seals. Seals are one of polar bears' main food sources. Seals swim in the ocean but breathe air. They cut air holes in the ice. Every few minutes, they come up to breathe. Polar bears wait by these holes. They catch seals coming up for a breath. As sea ice disappears, polar bears have a harder time hunting seals.

A warming climate is making food sources harder to find for land animals too. For example, caribou in North America migrate long distances in the spring. They give birth in areas with enough shrubs for female caribou to eat and raise their calves. However, due to a warming climate, the plants are done growing before the caribou arrive. There is not enough food for the large herds.

Some migratory animals such as caribou and moose are traveling farther and farther north. They are looking for cooler weather. Moose have been found in locations where they have never been seen before. When animals move to a different habitat, they create a ripple effect. The living things found naturally in a habitat have

Caribou can travel in very large herds. It is important that their habitat has enough food for all of them.

SEA TURTLE NESTS

Sea turtles nest on beaches. Many of these beaches are on islands. For example, researchers estimate that sea turtles use 30 percent of Caribbean beaches for nesting. When sea levels rise quickly, the water washes beaches away. Coastlines cannot adapt. Without enough space, sea turtles cannot lay their eggs. Sea turtles face other threats too. They are at risk of getting caught in fishing nets and being caught for food.

adapted to each other. They have defenses to make sure the population of each species stays balanced. When an animal moves to a different habitat, the species there might be defenseless against it. The new animal could eat a native plant or animal until it dies out. Scientists say that half of all animal species around the world are moving to new areas. The animals need to find suitable food sources.

Other migratory animals such as birds have a similar problem. Studies show birds are flying farther and for more days than they used to during migration. They

When a baby red knot does not get the nutrients it needs to grow, it may not be able to eat the food it needs as an adult.

are searching for places that are the right temperature at which to breed or rest over the winter. Birds stop migrating when they find the right temperature.

Birds with long migrations might not get enough food when they arrive. For example, the red knot is a bird that migrates from tropical areas to the Arctic. Red knots give birth to and raise their young in the Arctic. In the past, the birds fed on the insects that were hatching near shorelines. But insects are hatching earlier due to climate change. Red knots do not get enough food to help them grow. As a result, young birds' beaks do not grow to full size. When the young birds fly south, their beaks are too small to eat their main food source in their wintering grounds.

OCEAN HEALTH

The ocean holds most of the extra carbon dioxide that humans create. Warmer seawater allows more carbon dioxide to dissolve into the water. This puts the ocean out of balance. As more carbon dioxide enters the water, there is less oxygen. Additionally, the extra carbon dioxide makes the ocean more acidic. Many marine animals have a difficult time in acidic waters.

Many ocean habitats rely on coral reefs. One of the best-known reefs is the Great Barrier Reef off the coast of Australia. Corals are sensitive to temperature changes. They are also sensitive to the acidity of the ocean. As the ocean gets warmer, corals die.

Corals are animals. They live alongside a type of algae. Corals and algae work together. Algae produce food for the coral. Corals provide a home for the algae.

Bleaching puts coral reefs at risk of dying.

But when water temperatures get too warm, the algae leave.

Bleaching happens when algae leave coral. The algae give coral its color. When the algae are gone, all that is left of the coral is a light-colored skeleton. If the algae do not come back, the coral dies.

Even small temperature changes can be harmful. Corals are okay if the water temperature rises or drops by 3.6 degrees Fahrenheit (2°C) for a couple of days. But it is important that the temperature goes back to normal. If the temperature stays too warm or cold for too long, corals begin to bleach.

Coral health is important for many species of marine animals. Twenty-five percent of

> # PERSPECTIVES
> ## CHANGING CONSERVATION
> Nikhil Advani is a scientist with the World Wildlife Fund. In an interview, Advani said that most conservation programs do not consider climate change, but they should:
>> *Conservation biology has traditionally focused on historic threats to species, like habitat destruction and overexploitation. And while addressing those threats remains [important], it's becoming increasingly clear that we need to understand how climate change could harm the various species we're trying to protect.*

marine life makes its home in reefs. Fish, crabs, worms, and sharks all use coral reefs for shelter and food. Without reefs, many species would lose their habitat.

Ecosystems rely on balance. When environments shift due to climate change, plants and animals may have a hard time surviving. If climate change isn't stopped, plants and animals will continue to struggle to adapt to the changing planet.

FURTHER EVIDENCE

Chapter Three focuses on the effects climate change has on plants and animals. What is the main point of this chapter? What key evidence supports this point? Go to the article about ecosystems at the website below. Find a quote from the website that supports the chapter's main point.

PLANTS, ANIMALS, AND ECOSYSTEMS

abdocorelibrary.com/effects-of-climate-change

CHAPTER FOUR

STRONGER STORMS

Climate change is complex. Warming temperatures change weather patterns. As climate change progresses, serious weather events will become more frequent and more dangerous.

HURRICANES

Water vapor is a greenhouse gas. Warm air holds more water vapor than cool air. When warm air holds a lot of moisture, it does not cool easily. It also moves more slowly than

Climate change causes more severe hurricanes.

PERSPECTIVES

WHERE DOES THE WATER GO?

Hurricane Harvey dumped 60 inches (150 cm) of rain near Houston, Texas, in 2017. The heavy rainfall caused widespread flooding. But scientists believed the city itself played a part in the severity of the flooding. Compared to the countryside, cities have less open ground that can absorb water. Most of the ground is covered in roads, parking lots, or buildings. The water has nowhere to go. It collects in the streets. Researchers concluded that these factors increased the chance of extreme flooding in Houston by 21 times.

cool air. Air that is warm and moist can produce more rain showers than cool air. It is also the key to strong storms.

Warm, moist air becomes unstable when it meets cooler air. The unstable air forms clouds. Sometimes the air begins to swirl. Rotation is the start of hurricanes.

Scientists are still learning about the role climate change plays in hurricanes.

CATEGORY 5 HURRICANES BY DECADE

Decade	Count
1970s	🌀🌀🌀
1980s	🌀🌀🌀
1990s	🌀🌀
2000s	🌀🌀🌀🌀🌀🌀🌀🌀
2010s	🌀🌀🌀🌀🌀🌀

This infographic shows the number of category 5 hurricanes from 1970 through 2019 in the Atlantic. What do you notice about the graph? How does the infographic help you better understand the text?

But several studies have shown a connection between warm ocean water and hurricane severity. Researchers believe that when storms do happen, they are stronger. Warmer ocean temperatures produce stronger storms. For example, category 5 hurricanes are becoming more frequent. Hurricane strength is measured in categories from category 1 (the weakest) to category 5 (the strongest). There were more category 5 hurricanes in

> **SNOWFALL**
>
> In 2015 a huge snowstorm passed over the northeastern United States. The storm left 3 feet (0.9 m) of snow on the region. Cities such as New York City, New York, and Boston, Massachusetts, stopped public transportation. Thousands of airline flights were cancelled. Most major highways were not cleared for two days. Additionally, the blizzard caused a storm surge that flooded coastal towns in Massachusetts.

the Atlantic between 2000 and 2009 than in any other decade.

WHAT CLIMATE CHANGE MEANS FOR SNOW

Large snowstorms on the East Coast in the United States may become more frequent because of climate change. Warm air holds more moisture. Moisture can take the form of rain or snow. When warm air from the south meets cold air from the northeast, storms form. Large snowstorms can damage homes and cause power outages.

More snowfall isn't the only cold-weather effect of climate change. Polar vortex events are becoming

A polar vortex can cause ice to build up on bridges.

more common too. A polar vortex is the area of low air pressure and freezing temperatures found at both of Earth's poles. This combination can cause high winds. In recent years, polar vortex events have broken away from the Arctic and moved into southern Canada and the United States. Scientists believe climate change may be one factor behind polar vortexes moving to the United States. The Arctic is getting warmer. The warmer

Arctic air pushes cold air south. The polar vortex causes dangerously low temperatures. Water pipes in houses can freeze and burst. Cars might not start.

WHAT CAN BE DONE?

One of the ways to slow climate change is to change laws. Laws can encourage people to use renewable energy, such as sunlight and wind, more than fossil fuels. Scientists say most energy needs to come from renewables before 2035 in order to help stop climate change.

Renewable energy has its weaknesses. For example, wind and solar energy rely on those resources being available. Some places get little wind or sun. Also, sunlight and wind can charge up batteries. But the batteries do not last very long. This becomes a problem when there is a period of rainy or calm days. Researchers are working to make renewable energy more reliable.

Some businesses run wind farms to collect a lot of energy from the wind.

Additionally, some people are concerned over how renewable sources are collected. For example, wind turbines are loud. Many wind farms are located in rural areas. There is little research on how wind farms affect wildlife.

If laws create more demand for renewable energy, there will be more motivation and funding for scientists to do more research. They can find ways to make renewable energy efficient with less impact on wildlife. Scientists, politicians, and everyday people need to work toward using renewable energy if we are to stop climate change.

STRAIGHT TO THE SOURCE

Earth's temperature changes naturally. Some years it is warmer. Some years it is cooler. Climate scientist Kevin Trenberth explained in an interview what happens when Earth is warming naturally while also experiencing global warming:

> There is a large level of natural variability, but it is when the natural variability is going in the same direction as climate change effects that we suddenly break records. We cross thresholds and record new extremes. A small effect can translate into a large impact under those conditions. It's the straw that breaks the camel's back.
>
> Source: David Biello. "What Role Does Climate Change Play in Tornadoes?" *Scientific American*, 21 May 2013, scientificamerican.com. Accessed 31 Oct. 2019.

WHAT'S THE BIG IDEA?

Read this passage carefully. What is the main connection being made between natural changes in climate and changes driven by climate change? Name two or three details that support this point.

FAST FACTS

- Climate change has consequences that affect the whole world.

- Melting glacial land ice causes sea levels to rise.

- Melting ice reveals dark rock and sea water. The dark colors absorb heat instead of reflecting it. Shrinking sea ice results in habitat loss.

- Animals are affected as habitats and migration patterns change. Plants are blooming earlier in the season. Caribou are moving farther north. Birds are migrating more miles to reach suitable nesting grounds. Polar bears are losing their hunting grounds as ice sheets melt.

- Melting permafrost releases methane, a greenhouse gas.

- Warm, moist air becomes unstable when it meets cool air. Stronger storms are possible. The decade of 2000 to 2009 had more category 5 hurricanes than any other decade. Warmer temperatures create conditions that make fires more likely. Stronger rainstorms in cities increase the chances of flooding.

- Changing laws and studying renewable resources can slow the effects of climate change.

STOP AND THINK

Dig Deeper

After reading this book, what questions do you still have about the effects of climate change? With an adult's help, find a few reliable sources that can help you answer your questions. Write a paragraph about what you learned.

Say What?

Studying climate change can mean learning a lot of new vocabulary. Find five words in this book you've never heard before. Use a dictionary to find out what they mean. Then write the meanings in your own words, and use each word in a new sentence.

Take a Stand

Switching to using renewable energy is an important step in stopping climate change. But renewable energy has its weaknesses. What do you think people can do to encourage more research to help improve renewable energy?

You Are There

Chapter Two of this book discusses melting ice in the Arctic. Imagine you are on an Arctic expedition. Write a letter home telling your friends what you have found. What do you notice about the landscape around you? Be sure to include plenty of details.

GLOSSARY

adapt
to change in order to live in a certain environment

asthma
a medical condition that makes it hard to breathe

conservation
the act of protecting animals and plants

drought
a period of time when little rain or snow falls

ecosystem
a collection of living and nonliving things and their interactions

glacier
a large mass of ice and snow on land

habitat
the area where an animal or plant lives

migrate
to travel routinely from one place to another

storm surge
the rising of the sea due to changes in wind and weather patterns during a storm

vortex
an area of rotating air that stays in one place

ONLINE RESOURCES

To learn more about the effects of climate change, visit our free resource websites below.

Visit **abdocorelibrary.com** or scan this QR code for free Common Core resources for teachers and students, including vetted activities, multimedia, and booklinks, for deeper subject comprehension.

Visit **abdobooklinks.com** or scan this QR code for free additional online weblinks for further learning. These links are routinely monitored and updated to provide the most current information available.

LEARN MORE

Alkire, Jessie. *Harp Seals*. Abdo Publishing, 2019.

London, Martha. *How Climate Change Works*. Abdo Publishing, 2021.

Watts, Pam. *Ocean Ecosystems*. Abdo Publishing, 2016.

INDEX

Advani, Nikhil, 30
Africa, 9, 10

Canada, 17, 21, 37
caribou, 24–25
coral, 28–31

drought, 9, 10

Environmental Protection Agency, 7
Europe, 6–7

fires, 6, 7, 9–10
floods, 10, 17, 34, 36
fossil fuels, 7–8, 11, 38
France, 5

glaciers, 8–9, 13, 14, 15, 18, 21

hurricanes, 34–36

islands, 15–17, 26

laws, 11, 38, 40

moose, 25

Ok, 14

permafrost, 20–21
polar bears, 24
polar vortex, 36–38

red knots, 27
renewable energy, 38–40

sea ice, 8, 13–15, 18, 23–24
sea levels, 15–17, 19, 21, 26
sea turtles, 26
seals, 24
snow, 13, 18, 36
Spain, 6

Thunberg, Greta, 11
Trenberth, Kevin, 41

World Wildlife Fund, 30

About the Author

Martha London lives in Minnesota and writes books for young readers. When she isn't writing, you can find her hiking in the woods.